Movies On Paper Studio

Presents

Scientist Rain

Title Page for: **Scientist Rain**

These entire children's book collection and series were created By: Dionne L. Fields

Dionne L. Fields also made other contributions to this book title.

Illustrated

Photograph

Proofread

Editorial

Written

By Dionne Fields
1. Fields, Dionne. -Business 2.Woman Business Owner 3.African American Author Patients-United States-Biography. 4. Publisher and Entrepreneur.5. Movies On Paper Studio.
6.Fields, Rain. 7. Ideas On Paper, 8. Sitcoms On Paper, Episode 1 and 10. Episode 2.

Book Title: Scientist Rain
ISBN-13:978-1507610749
ISBN-10:1507610742
 By Dionne L. Fields On January 14, 2015

Movies on Paper Studio **NonFiction** Children's Book Collection.

1.Rain's 1st Christmas,
2.Actor Rain,
3. Inventor Rain,
4. Super Model Rain,
5.God's Child One
6.Recording Artist Rain,
7.Surfboarder Rain
8. Fashion Model Rain,
9. President Rain,
10. God's Child Two,
11.CEO Rain,
12 Author Rain
13. Bully Proof Rain
14.Rain Boy Fashion,
15. Football Team Rain,
16. Business Mogul Rain,
17. Pilot Rain,
18. Attorney Rain,
19. Rain's Children's Library,
20.Super Rain,
21. Philanthropist Rain,
22. Poet Rain
23.Rain Storybook Poem
24.Happy Mothers' Day.
25. Rain Magical Library
26. Rain Fields Incorporation
27.Rain Retirement Paorty,
28.Sir Knight Rain
29. Rain Vacation
30. Rain' Famous Friends

31. Fire-fighter Rain,
32. Prince Rain,
33. Musician Rain,
34. Fashion Designer Rain,
35. Cupcakes By Rain,
36. Astronaut Rain,
37. Photographer Rain,
38. Chef Rain,
39.Rain's Children's Book Museum,
40. Race Car Driver Rain,
41.The Rain Fields Children's book collection,
42.Atlantis Rain,
43.Sextillion Dollar Rain,
44.G.I.Rain,
45.Scientist Rain,
47.

Movies on Paper Studio **Fiction** Children's Book Collection.

1.BlueBerry Bedtime Story,
2. The Magic Cell Phone,
3.Red Rain Boots,
4. Bubble Bath Time,
5.Broccoli Meet Cheese,
6.Good Night Blanket,
7.The Lost Tooth,
8.Read Fairy,
9. Sitcoms On Paper, Episode 1
10.Episode 2

Movies on Paper Studio NonFiction Novels.

1. Movies On Paper Studio,
2. Dionne Fields Incorporation,
3. Honoring Colleges,
5. Documentary Of Real Champ,
6. Living For Today
7. September 11,
8. Sedreck Fields,
9. Greatness Endured 37 Losses,
10. Documentary Real Champ, 11. Sedreck Fields Scholarship Fund, 12. TenThousand Volunteer Hours, 13. No One Exempt From Hard Times, 14. National Book Signing, 15. Black History & Me, 16. 2012 By Dionne, 17. Theresita. 18. Unlimited God favor, 19. Black Movies On Paper, 20. Sedreck Fields Foundation, 21. Purpose & Reflections, 22. Behind Close Doors, 23. Facing Tomorrow 24. One Wish 25. History Maker Nominee 26. Dionne Fields Reality TV Show, 27. Cancer survival museum, 28. Mascara, 29. The Director, 30. Waverly, 31. The Girl's Club. 32.

Movies on Paper Studio True - Crime Book Series.

1. The Untold Story,
2. Unpunished,
3. Whispering A Secret,
4. Pages Of Me Chapter One,
5. Pages Of Me Chapter Two,
6. Pages Of Me Chapter Three,
7. Pages Of Me Chapter Four,
8. Pages Of Me Chapter Five,
9. Pages Of Me Chapter Six,
10. Pages Of Me Chapter Seven,
11.

Movies on Paper Studio Ideas On Paper Book Series.

1. Re-designed Living Room Suite 2. Re-designed Lightweight Military Uniforms 3. Re-designed Toddler Strollers 4. Re-designed Ball Game Seats 5. Re-designed Lamps 6. Re-designed Bathroom Sets 7. Re-designed Business Chairs 8. Re-designed Car Seats, 9. Re-designed wall covers

This children's book collection was inspired, by the real life character "Rain Fields".
One of America's favorite storybook characters of today.

First published in the United States of America in 2015
By Dionne Fields (Publisher) new hardcover edition.

Library Of Congress Cataloging-In-Publications Data
Fields, Dionne L.

[Movies On Paper Studio, Coming Soon To A Shelf Near You.]

Originally Published, 6331 Pleasant Ridge Road. Knoxville, Tennessee.37921

Library Of Congress Control Number:
ISBN-13:978-1507610749
ISBN-10:1507610742
http://www.onlineslibrary.webs.com

This novel was written and created By Dionne Fields

Printed in the United States of America January 14, 2015

Content Page

Movies On Paper Studio

Scientist Rain

Introduction by, Dionne Fields

Introduction page

Complete Book Title

Scientist Rain

Chapter 1

The story book about a young boy, who wants to learn about becoming a Marine Biologist (scientist).

Chapter 2

The complete Rain Fields children's book collection.

All of Rain's Limited Edition Books, are already Autograph.

To My Friends & Book Fans

From

Rain Fields

RAIN

Chapter
One

Scientist Rain

Elementary Kids choosing careers paths early in life.

I want to inspire kids to learn and study more about what they would love to do when they grow up.

This storybook is to encourage kids of all ages to read more and to study, their career choice early on.

When they enter high school, they would have learn more about their career choices and what path to work towards and college choices.

Rain a little boy in elementary school wants to learn more about becoming a scientist when
He gradates from high school .

He is on a path about learning, to be come a marine scientist, someday.

Rain has been working on his career since may 2012.

He wanted to share this children's book with other school-age kids, to inspire them to also read more and learn about your career choices early.

Most major industries have, tours and times to go look at some career choices earlier with your parents.

I wanted to write a storybook for children, to inspire them with the tools they may need to help them choose the right career for them to learn and study early.

In chapter two, in this book you will see some examples of Rain's Career choices that he has made every since he could walk.

The information in this book, was gain by reading and studying other books at his local elementary school library.

Scientist Rain

A person who studies ocean life are named a **Marine Biologist**.

Oceans cover more than two thirds of the earth's surface, and they are deep.

Scientist Rain

Rain has read in many books. That the ocean, are filled with many plants.
Which provide food, homes, and protection for ocean animals.
Fish and other animals make their homes in all different parts of the ocean, even on the
bottom, or seafloor.

Scientist Rain

This storybook is to encourage kids of all ages to read more and to study, their career choice early on.

When the have enter high school, they would have learn more about their career choices and what path to work towards and college choices.

Scientist Rain

Rain enjoyed learning about becoming a marine scientist .

He enjoyed reading all about fishes, in the ocean world from his school library.

Scientist Rain

Rain enjoyed using his own camera, that he had received from his grand mom on his 5th birthday .

He had fun taking pictures of all types fishes that lives in the ocean.

Scientist Rain

Rain has learned that fish have bones and are cold-blooded. Cold-blooded means that their body temperature depends on the temperature of the water around them, unlike humans, who have a normal body temperature that stays at 98.6 degrees.

06.08.2012

Scientist Rain

Rain has learned that most fish have a special part inside their bodies called a swim bladder.
Air inside the swim bladder helps the fish float through the water.

Scientist Rain

Most fish can breath oxygen from the water through gills on the sides of their bodies. They cannot breath air or live outside water.

Scientist Rain

Rain has learned that fish are covered with small scales instead of skin

06.08.2012

Scientist Rain

Rain has learned that some microscopic plants and animals called plankton are important in the ocean because they become food for many animals.

06.08.2012

Scientist Rain

Rain has read in a book at his elementary school library.
That a fish use fins and a tail to swim around.

06.08.2012

Scientist Rain

The was the most beautiful fish that Rain, has ever seen.

I'm so glad that Rain could take a photo of this beautiful pink and black fish.

Scientist Rain

Every day a group of fishes swims to school.

06.08.2012

Scientist Rain

This fish kept staring at me.

Scientist Rain

I could not get this fish to smile today.

Scientist Rain

This Red fin fish, is still following me around with my camera.

Scientist Rain

Since this fish has a personality, I'm going to name him Mr. Red Fin.
Because his fins are red

Scientist Rain

The bigger fish in this photo, looks like a teacher in their school.

And the other fishes, looks like students in school.

06.08.2012

Scientist Rain

This fish looks like the school bully.

06.08.2012

Scientist Rain

Rain had fun learning about all types of fishes .

And if he keeps learning more new creatures, than one day he could be the next marine scientist.

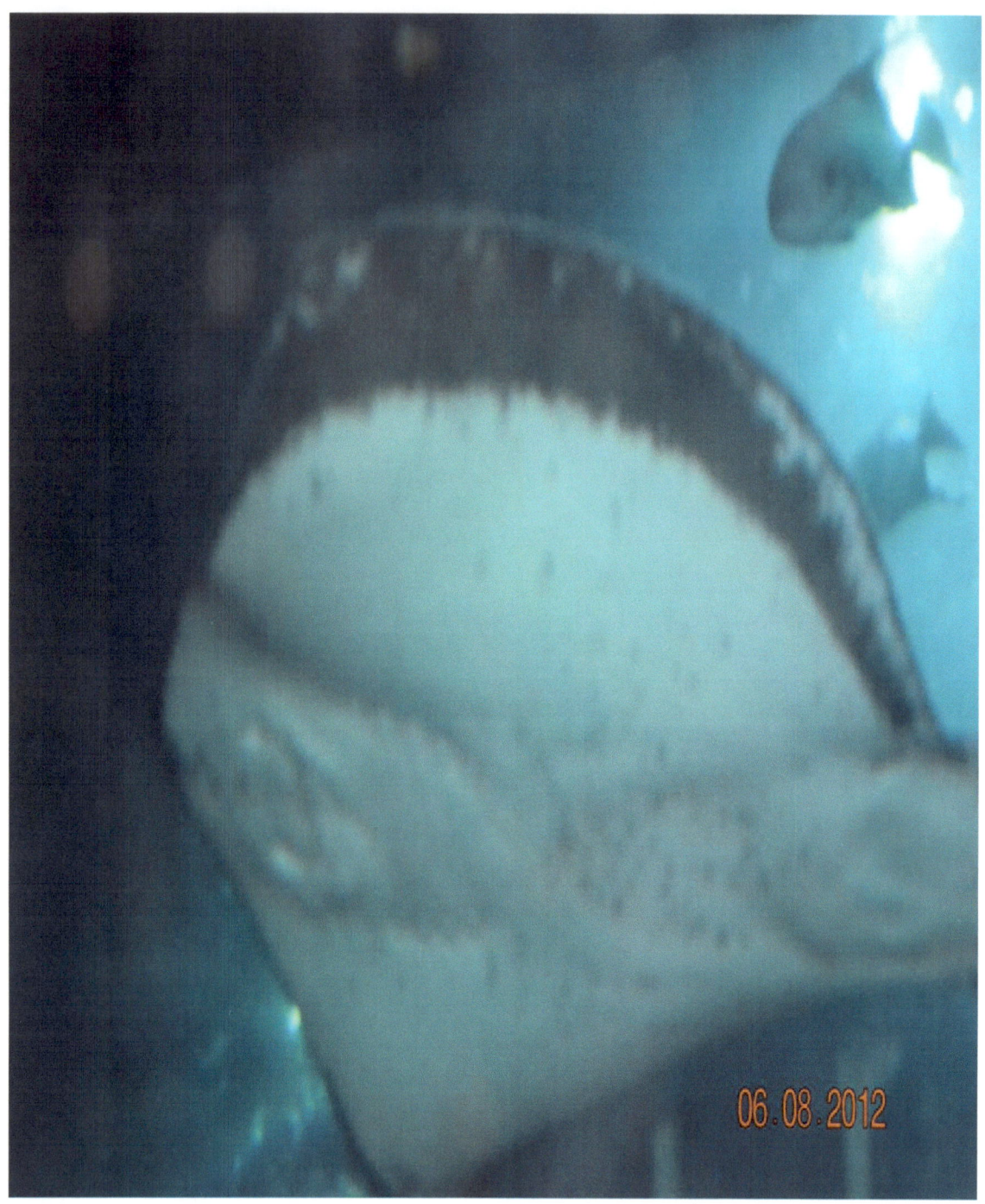

06.08.2012

Scientist Rain

A stingray look like flattened fish with eyes on top of their heads, gills underneath their bodies, and a long tail.

Rain has learned that some rays like to live alone, but most of them live in large groups with other rays. Some kinds of rays have spines on their tails that have poison, which they use to stun or kill their prey. Some rays eat like whales - they filter small pieces of food out of the water.

06.08.2012

Scientist Rain

Rain has learn that the **jellyfish are found in every ocean, from the surface to the deep sea.**

06.08.2012

Scientist Rain

The jellyfish was exciting for Rain to learn that **a gelatinous umbrella-shaped bell and trailing tentacles.**
 The bell can pulsate for locomotion, while stinging tentacles can be used to capture prey and their food.

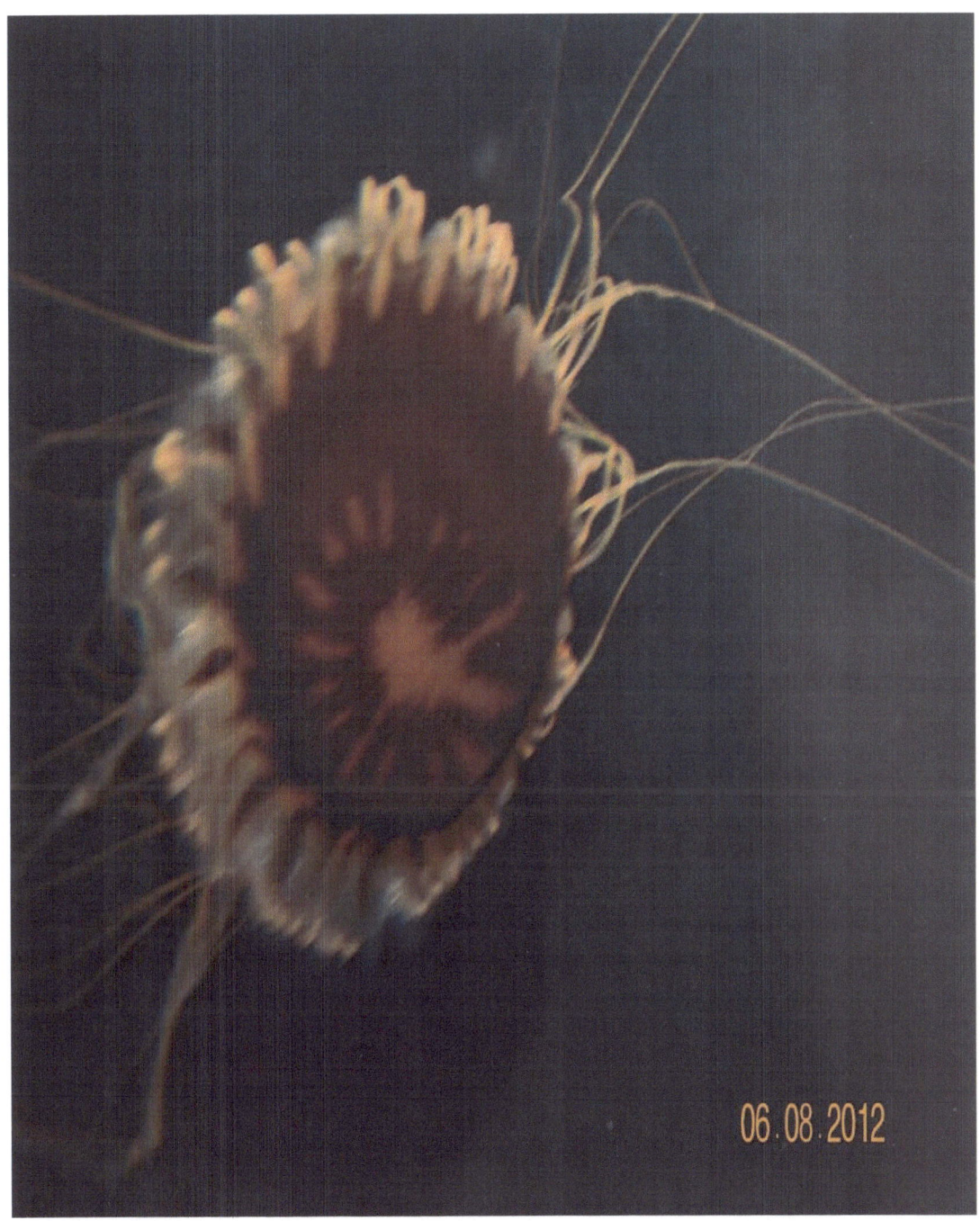

06.08.2012

Scientist Rain

A jellyfish are large and colorful, Jellyfish have roamed the seas for at least 500 million years.
 This would make a jellyfish the oldest multi-organ animal.

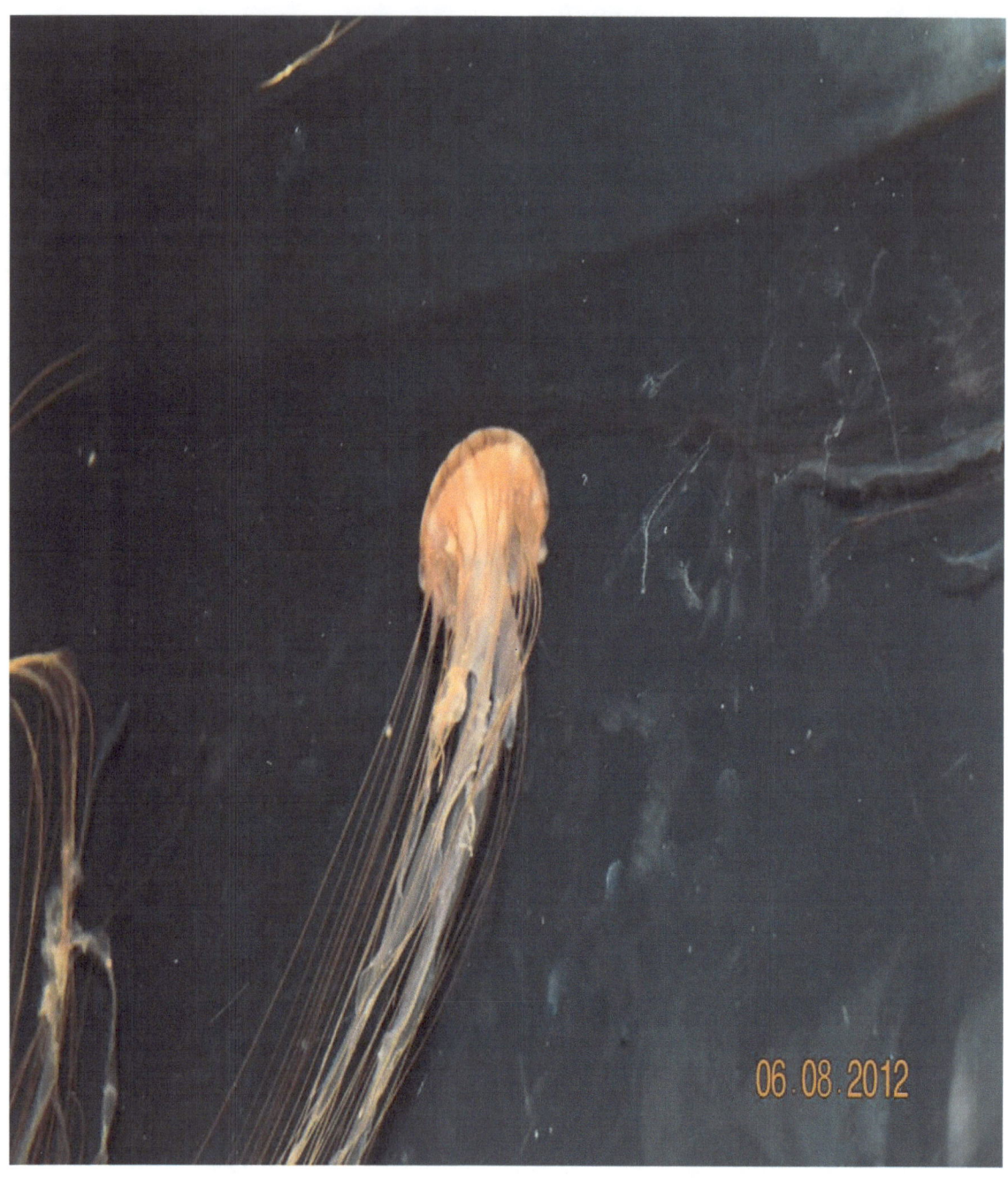

06.08.2012

Scientist Rain

Rain has learned that a jellyfish do not need a respiratory system.
Since their skin is thin enough that their body can breath oxygen.

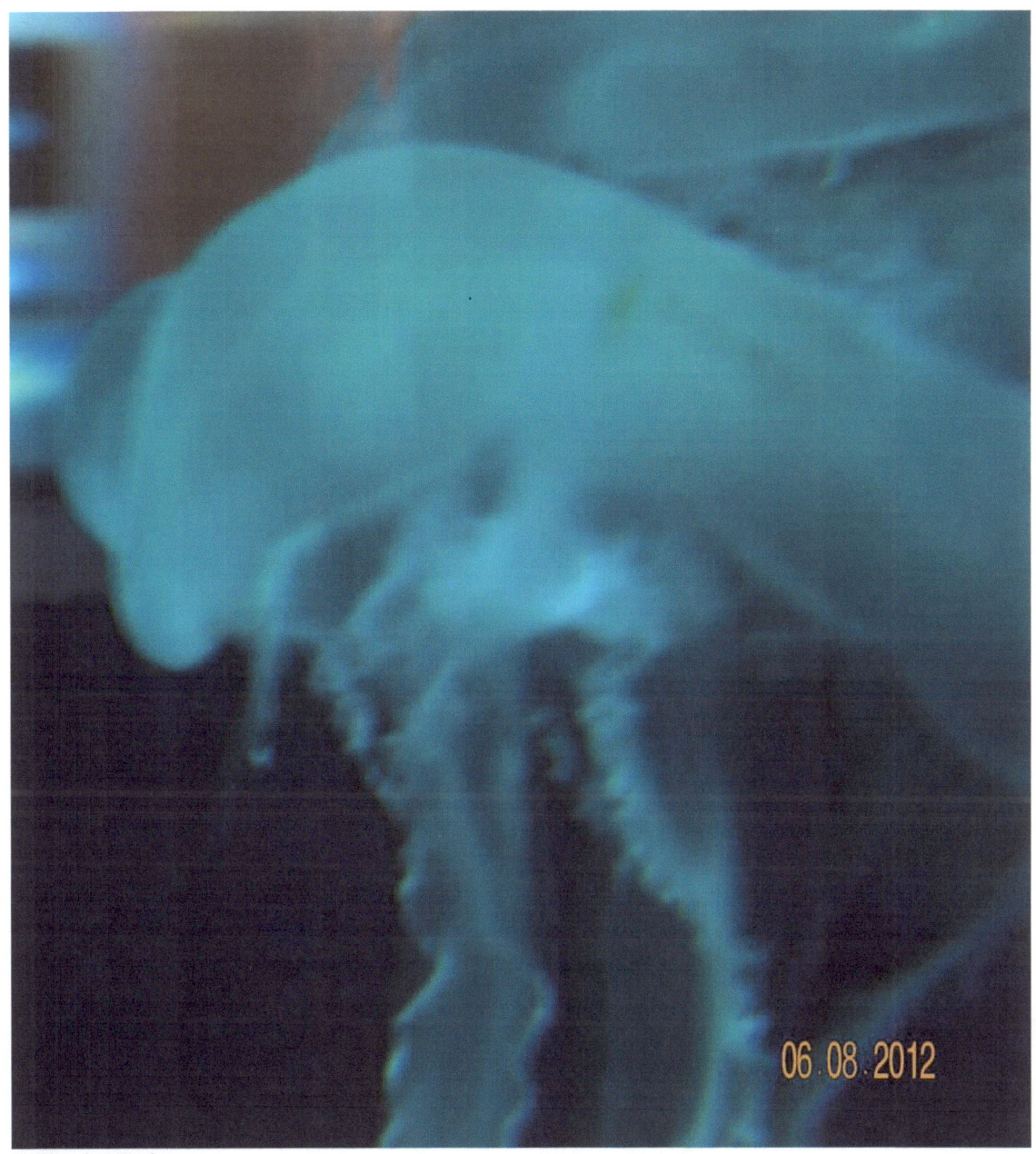

06.08.2012

Scientist Rain
A jellyfish remain in still waters until the tide rises.
They are swept back into the ocean bay near salt water.

06.08.2012

Scientist Rain
Rain has also learned that a jellyfish can detects all types of movements.
Including the touch of other animals in the ocean using their nerve cells.

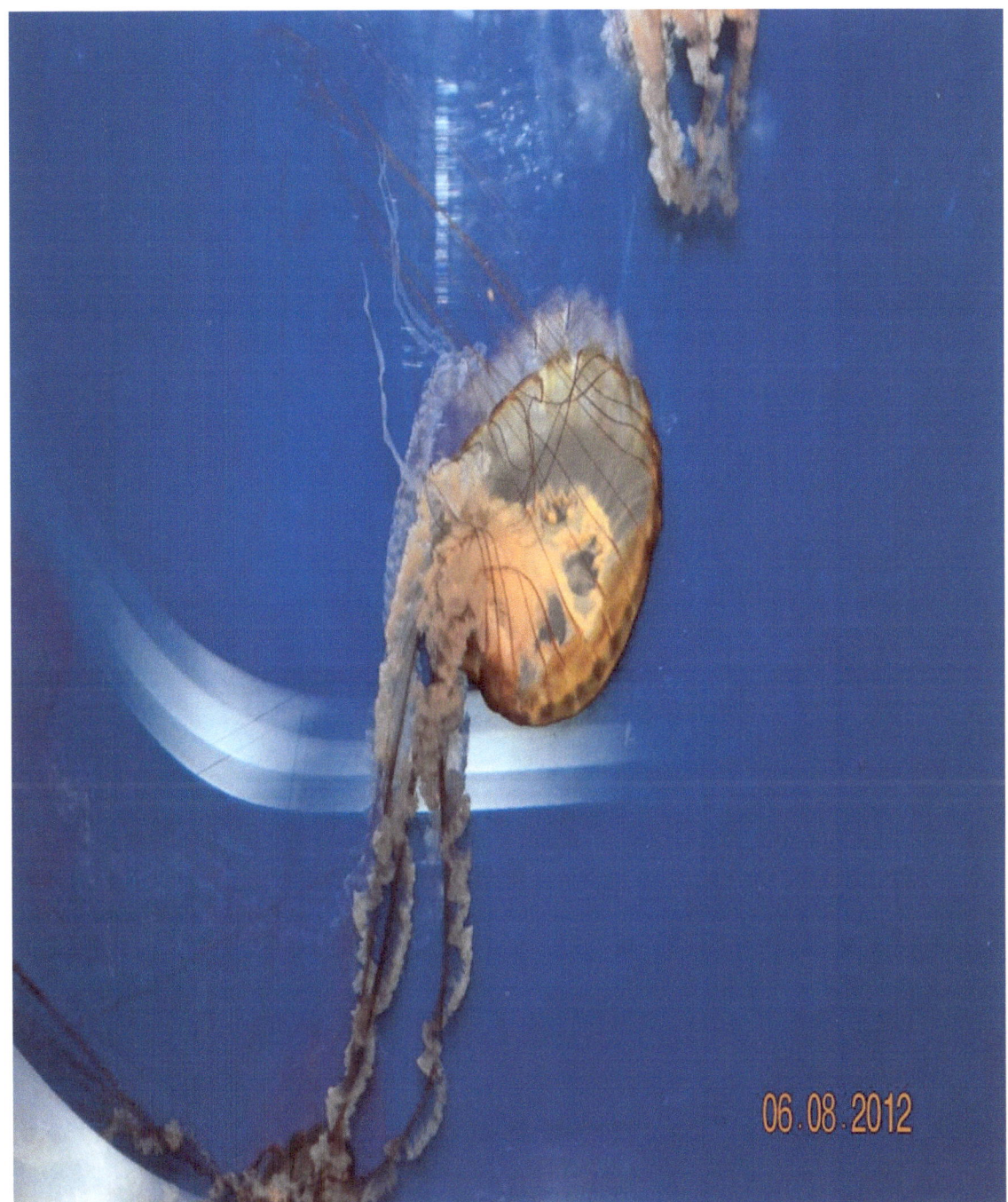

06.08.2012

Scientist Rain

Rain has learn today at his elementary school library, that a box jellyfish has twenty eyes. And two of the jellyfish eyes can see all types of color, and also have a 360-degree veiw of the oceon.

06.08.2012

Scientist Rain

Rain have seen pictures jellyfish in books at school.

He was very happy to see a jellyfish in real life.

Scientist Rain

Rain wanted to inspire kids of all ages to read books.

And to continue to check out books each week, at your school library.

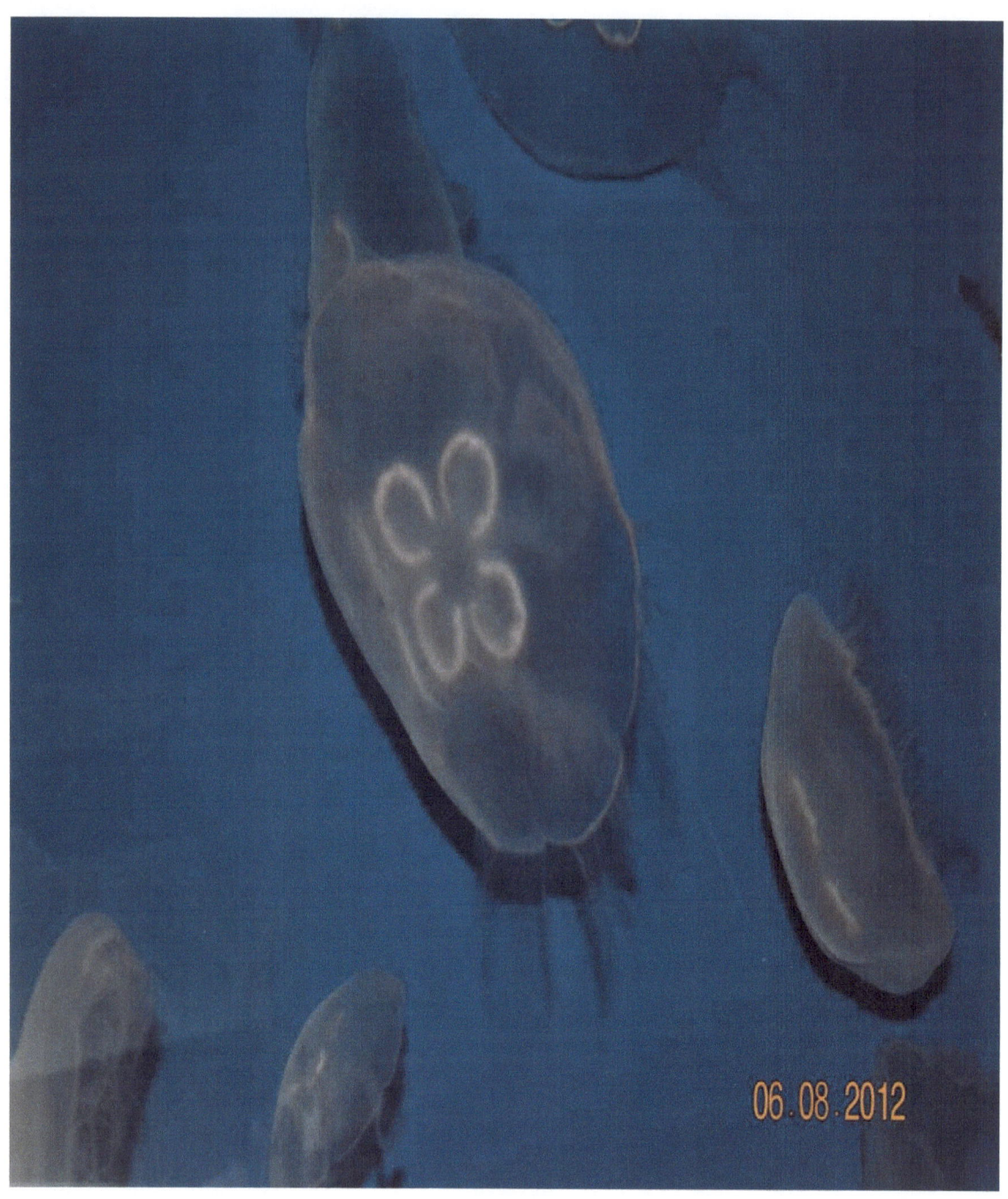

Scientist Rain
Rain has learn that the jellyfish can range from about one millimeter in bell height.
And diameter to nearly two meters in bell height and diameter.

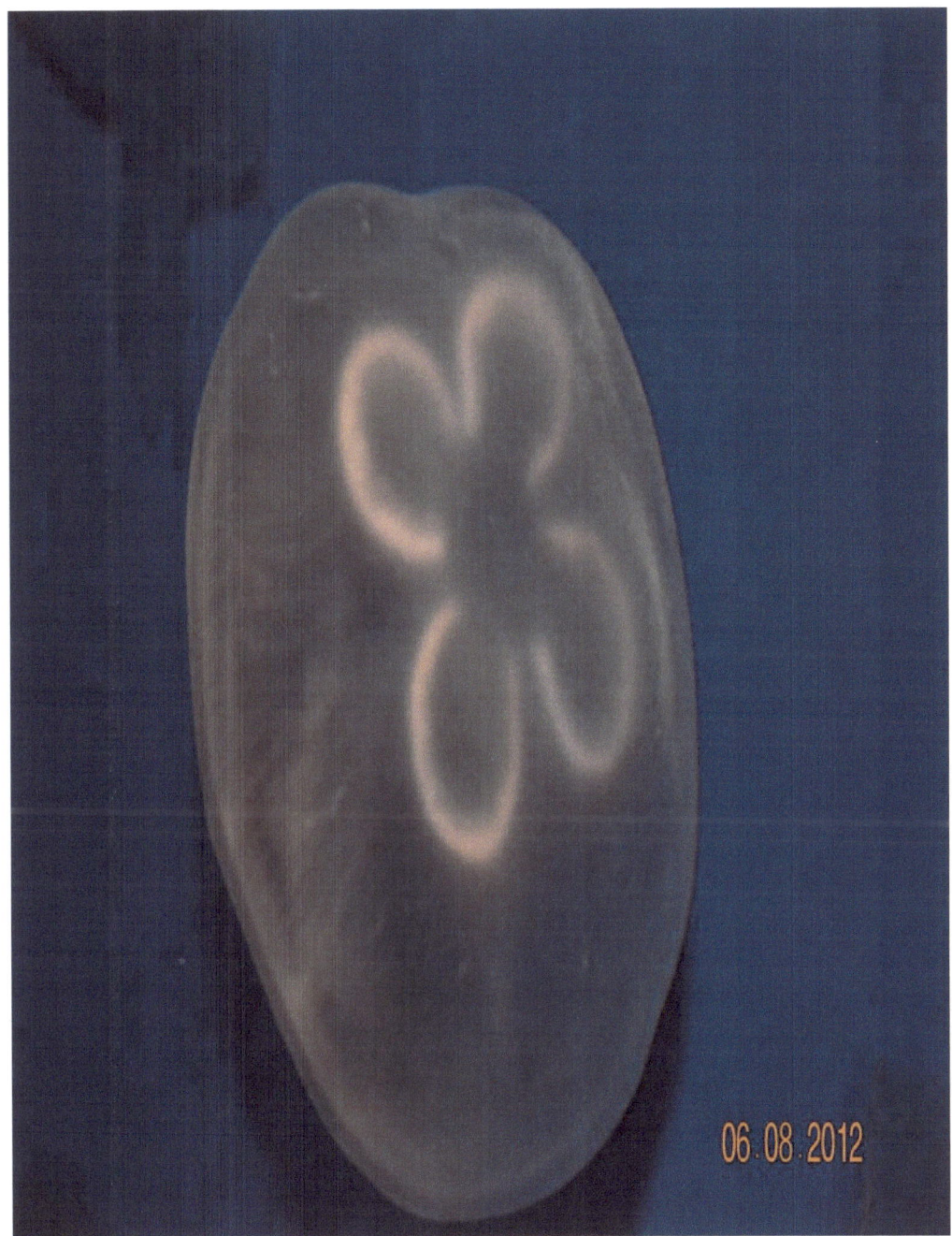

06.08.2012

Scientist Rain
Did you know that the smallest jellyfish are the peculiar creeping jellyfish?
In the which have bell disks from 0.5 mm to a few mm diameter with short tentacles.
A tiny creeping jellyfish cannot be seen without a hand lens or microscope.

Scientist Rain

Rain has learn that seahorses are found in shallow tropical and temperate waters around the world.

Scientist Rain

Rain has also learn from reading books at his school library on seahorses.

That a seahorse males stay within 1 m^2 (11 sq ft) of habitat,.

And the female seahorses range about one hundred times within habitat.

Scientist Rain
A Seahorse can range in size from 1.5 to 35.5 cm (0.6 to 14.0 in).
They are named for their equine appearance, they look like miniture horses.

Scientist Rain
A seahorse has a flexible, well-defined neck.
Rain has learn that seahorses swim upright.

06.08.2012

Scientist Rain

Rain has read in a book at his school library.

That a seahorse is the slowest-moving fish in the world.

06.08.2012

Scientist Rain

A seahorse do not have scales or fins.
They just have a thin skin stretched over a series of bony plates in their body.

Scientist Rain

Rain was happy to learn, that a seahorseuses their long snouts, to suck up food.

06.08.2012

Scientist Rain

The seahorses are poor swimmer in the oceon.

They use their tales, to hold on to other objects in the water.

Scientist Rain

A seahorse has eyes that can move independently of each other .

This helps them to look out for preys, since they can't swim very fast.

Scientist Rain

This solid white seahorse, was the most beautiful seahorse I ever seen.

Scientist Rain

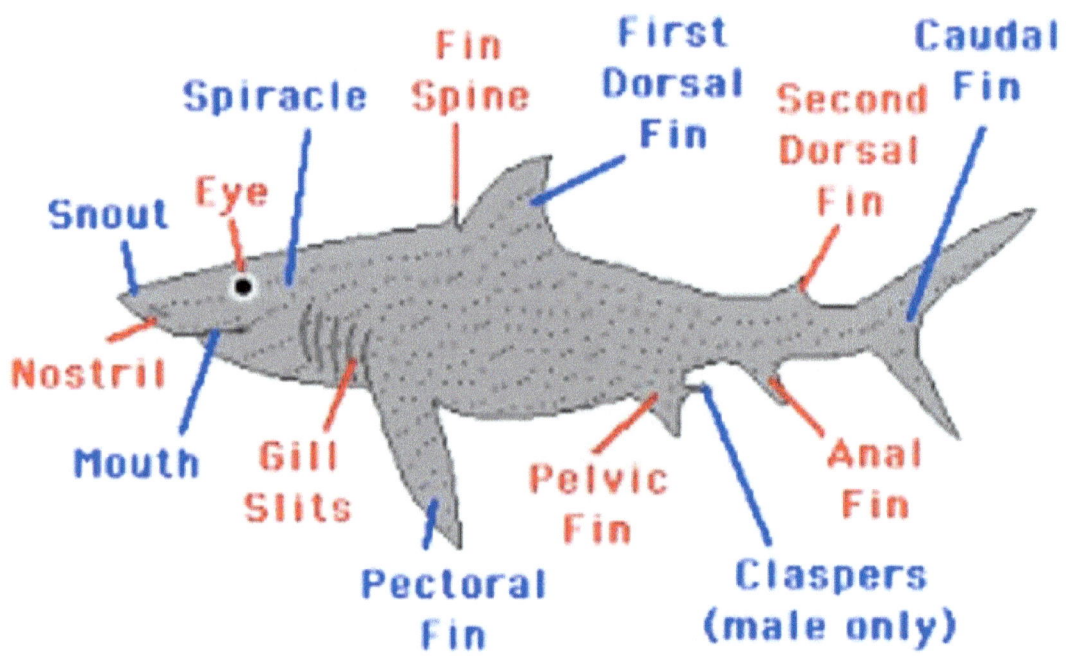

06.08.2012

Scientist Rain

Rain has learned that most sharks look a lot like marine mammals, but they are actually fish, sharks do not have bones.

06.08.2012

Scientist Rain

All of sharks skeletons are made up of cartilage instead of bones.
Cartilage is the same stuff that makes your ears and the tip of your nose have their shape
and be a little bit flexible.

06.08.2012

Scientist Rain

Rain had a lot of fun, learning a lot about becoming a marine scientist some day.

He enjoyed reading all about sharks, in the ocean world from his school library

06.08.2012

Scientist Rain

Rain was excited to learn that sharks have several rows of sharp pointed teeth and powerful jaws to tear off big pieces of food, which they swallow whole.

Scientist Rain

Rain loves visiting the library, each week at school with his class. This week he has
Been learning about sharks.

06.08.2012

Scientist Rain

Did you know that when sharks do attack people, it is probably because they mistake people for seals or other large ocean animals that they would like to eat.

It is hard for them to tell what's what from below the surface of the water.

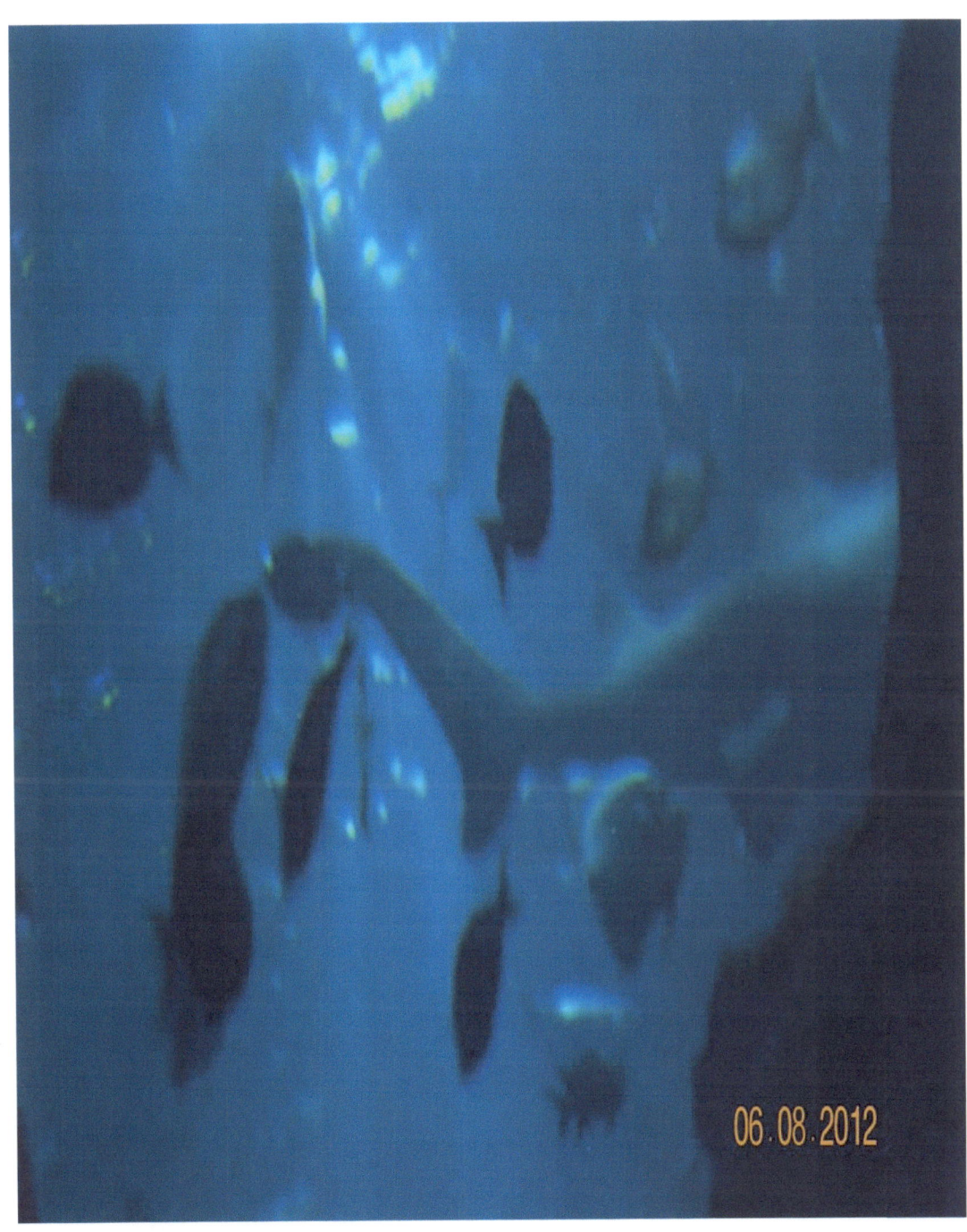

06.08.2012

Scientist Rain

Today Rain has learn that there are over 350 kinds of sharks, but only about 25 of those have ever been known to attack humans.

Scientist Rain

Rain also learn that some sharks only grow up to 7 inches long. However, most kinds of sharks grow to about 5-7 feet long, which is about the same height as an average adult

06.08.2012

Scientist Rain

Sharks have an excellent sense of smell, which helps them find food.

Many sharks eat near the surface of the water, but will also dive down deeper in search of food.

06.08.2012

Scientist Rain

Did you know? That some sharks they don't even need to chew their food! Sometimes sharks lose teeth, but when they do, new ones grow in their place.

Scientist Rain

Rain has learned, that some sharks has very long nose.

06.08.2012

Scientist Rain

Even though sharks don't have bones, sharks do have lots of other similarities to bony fish.

They are cold-blooded and breathe through gills instead of through lungs. Sharks have gills on each side of their heads.

Rain has also learned that as they swim, water passes over the gills and oxygen flows in from the water.

Scientist Rain

Rain was so surprise, that baby sharks are called pups.

Some pups grow inside their mothers like human babies do, but most hatch from eggs.

Chapter
Two

Scientist Rain

The beginning of chapter two.

In this section of the story book Scientist Rain, you will learn more about some of the career choices Rain has learned about.

I wanted to write a story book for children, to inspire them with the tools they may need to help them choose the right career for them to learn and study early.

In chapter two, in this book you will see some examples of Rain's Career choices that he has made every since he coluld walk

Rain had a lot of fun, over the past several years in his life choosing what type of career he wanted to do.

Author Rain Fields

Rain was bless to publish his first childrens book.

When Rain was only three months old, he wanted to read a book.

Attorney Rain Fields

He won his 1st case at the age of 1st months over chocolate chip cookies with milk.

This book is full of all the wonderful things a child does on a daily basic to win a kids case.

ASTRONAUT RAIN

Dionne L. Fields

Astronaut Rain Fields

Rain is very fasinated about the stars and moon at night.

He wants to become a astronaut and walk on the moon some day.

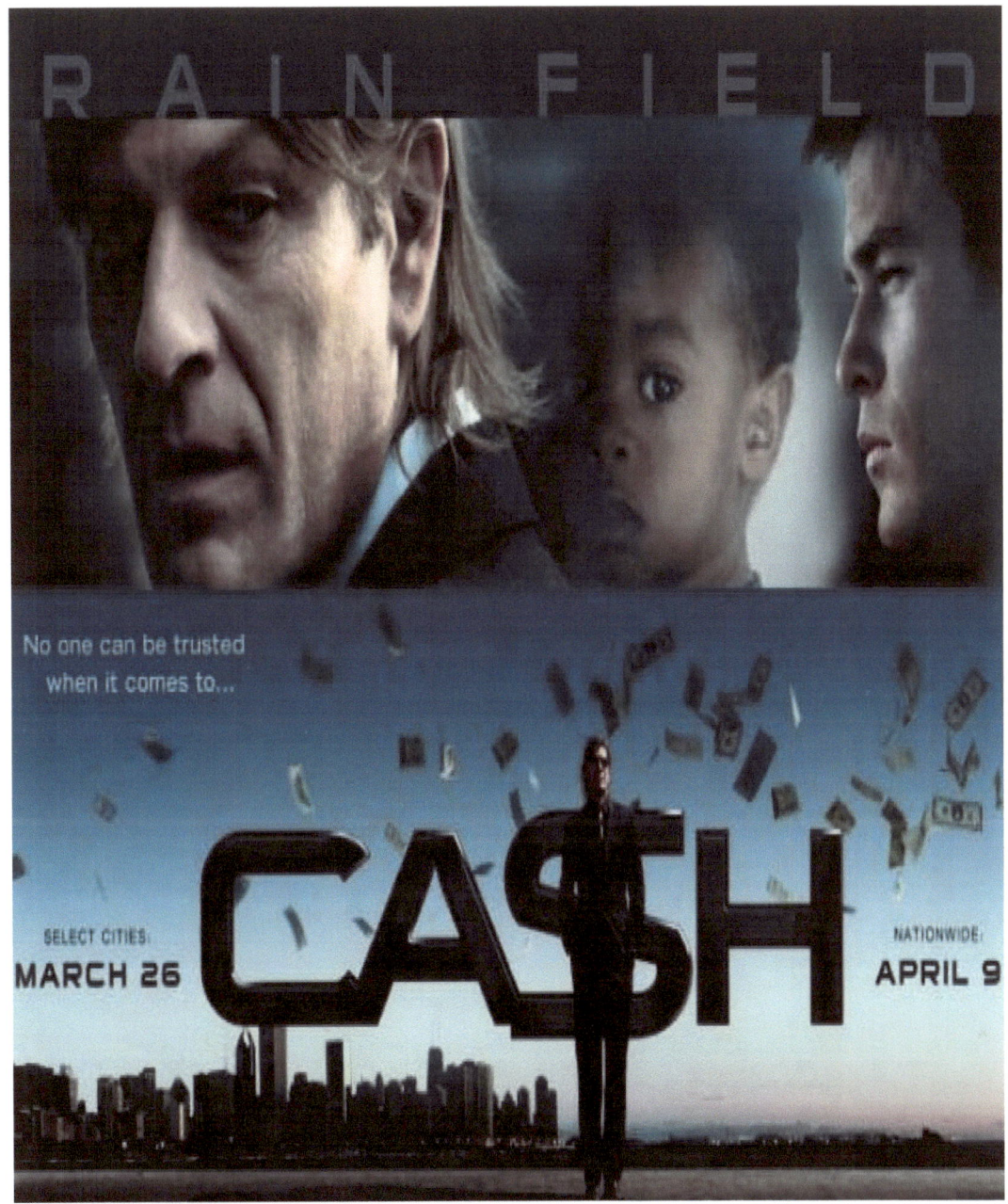

Actor Rain Fields

My mother has raised me to believe, that at any age.

I can be what ever I wanted to be in life.

I just need to work hard and make good grades in school.

I want to be a famous actor one day.

The Adventure Of Rain Fields

The Adventure Of Rain Fields, **Sitcoms On Paper**.

Episode 1 and Episode 2 will be available in the children's book section at online library & bookstore.

I wanted to give kids of all ages the opportunity to read each Sitcom On Paper, before it comes out on television

The Atlantis Rain

The Atlantis Rain:

The story is about a boy around the age of seven years old
A second grader, who fines learning about science and history in school
Could be a fun adventure.
While on a school field trip to the Smithsonian museum.
The boy discovers the myth about the Atlantis world on his own.

The bully proof system is my way, of helping student all across the world from being bullied at school.

This new Bully proof system will help student to be able to learn more.

And to not worry about being bullied.

Baby Fashion Model.
" Rain Fields "

Fashion Model Rain Fields

Fashion Model, Rain Fields all the funds from this book.

And from modeling will go towards me having a better future.

My Mother will make sure there is money for my education.

Rain scholarship Fund for college, money for me to attend a great christian school.

Rain's First Christmas

My 1st Christmas was really fun.

I wanted to share this beautiful moment to all the babies.

I hope their 1st Christmas was blessed like mines.

Cupcakes By Rain Fields

Rain loves baking cupcakes for all his friends and kids.

Rain's kid's cookbook of cupcakes is so fun and very good.

Carrot cupcakes, top with real shredded carrots on top, with Orange marmalade frosting.

Gladiataor Rain

The Gladiator Duel is the most popular battle game for kids.
The objective is to use our foam filled gladiator sticks, to joust your opponent off their podium and be declared the Gladiator Duel Champion.
Don't worry as when your opponent falls off they land on our inflatable safety cushion.

Scientist Rain

Rain a little boy in elemetary school wants to learn more about becoming a scientist when He gradutes from high school .

He is on a path about learning about being a marine scientist, he has been working on this project since may 2012

Ceo Rain Fields

Rain at 3 months old, his 1st day at work.
He just turns 8 years old on Mother's Day.
He is so excited about attending 3rd grade in the fall.
He is the ceo of his own business, with the help of his mommy(Dionne Fields).
Rain Fields has been very busy the last 4 years.
Rain Fields Inc. www.rfields.webs.com

Fire-Fighter Rain Fields

Rain wants to protect and serve all the kid's in his community.

http://www.youtube.com/watch?feature=endscreen&NR=1&v=6IvG1Bp7z-0

Football Team Owner, Rain Fields

I believe any kid can be, who ever they want to be in life, with hard work and a good education.

My hope and dreams one day to own my very own Football Team.

I would name my own Football Team Virginia Ballers, from the home state my mother was born.

Fashion Designer Rain

Rain loves designing new clothes for kids of all ages.

Rain has his very own clothing label , Rain Boy Fashions.

©Rain Boy Fashions February 14, 2014

Gods Child # 1

Gods Child # 1 is a children's book about how special children are to God. Every child is Gods child.

Rain knows he is special because of his mother's unconditional love for him.

Gods Child # 2 is a new line of my children's books.

My two sons, has inspired me.

To share with children, of all ages.

How fun life can be as a child of God

Happy Mothers Day.

Happy Mother day mommy.

Thanks for always believing in me.

That I could be whom ever, I dream of at any age.

Inventor Rain Fields

Inventor Rain Fields is a new child inventor.

Rain has invented a product for kids, to protect them from germs and bacteria.

I'm a blessing.

This children's book is from, the rain field's children's book collection.

A boy raised by a single mom, even in her daily struggles to provide for him.

She reminds him that he is her biggest blessing.

This wonderful book will put, a smile on every Childs face. And most of all every child is a blessing from God.

123 Bathroom Safety 4 me

All natural disinfectant spray

Rain Fields is very happy about the launch of his new kid's personal care line.

Jesus Loves Me.

A story about a little child's, faith in God.

Yes, God does love all children all across the world.

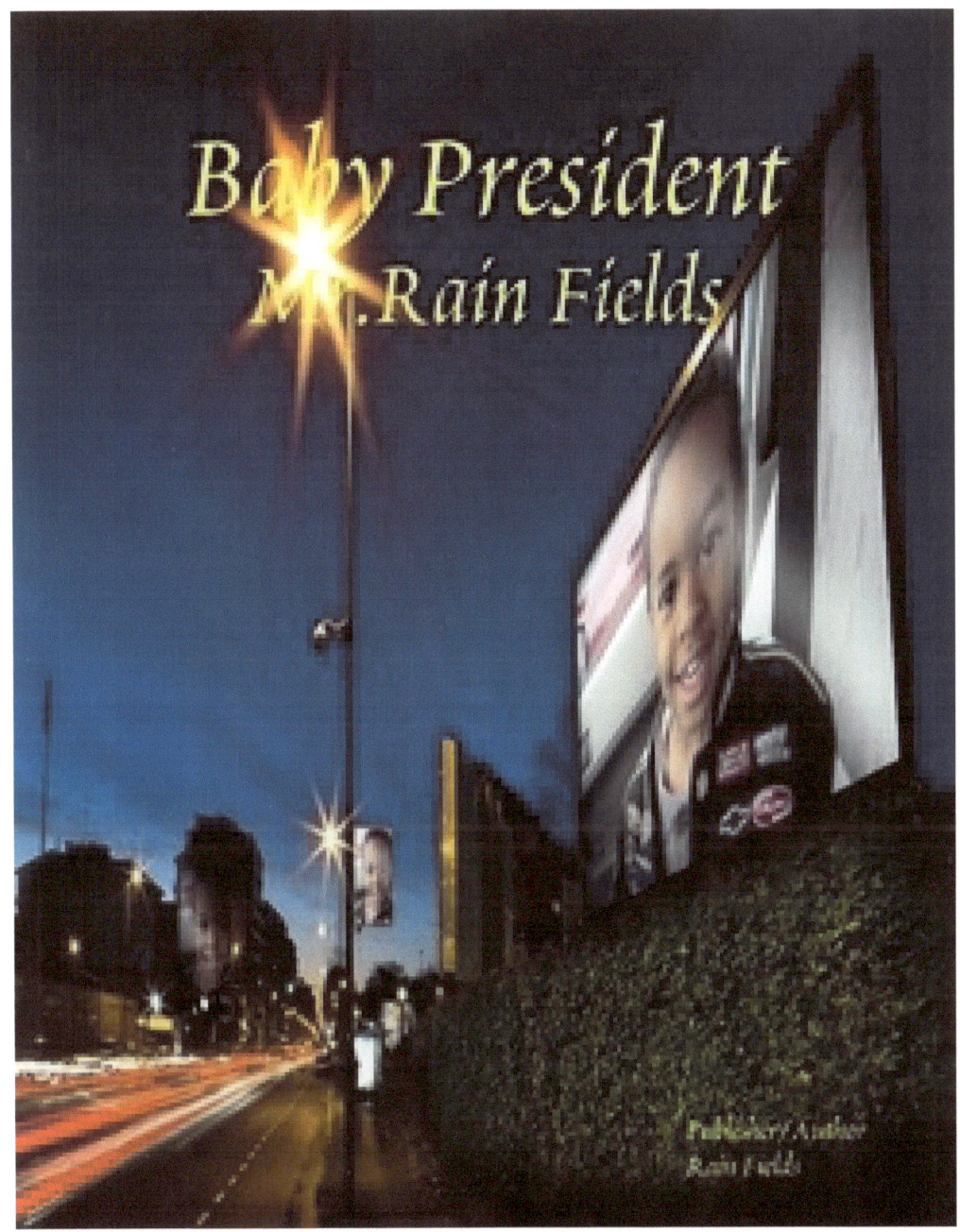

President Rain Fields

I would love to be President of United States Of America.

I would just be thrill to visit the white house in Washington Dc.

This children's book is to encourage all kids to do great in school, they could be the next President one day.

Philanthropist Rain Fields.

This foundation will help every single mom with babies or toddlers.

Rain will be giving away for free, yes there is a waiting list for certain baby items.

The Rain Fields Foundation is a monthly baby and toddler shower for single moms.

Poet Rain Fields

This is a storybook of poems for kids.

All the poems in this children's novel, are inspired by Rain Fields.

Poet Rain Fields, features a dozen fun poems for kids of all ages.

Prince Rain

For years girls of all ages, are raise to become future princess.

Now rain would love for all boys to dream of becoming a prince.

Rain dreams of becoming the next prince one day.

Lights
Camera
and
Action.

Photographer Rain

Rain always wanted his very own camera, to take lots of photos.

He is very happy about ,sharing some wonderful pictures to the world.

Pilot Rain Fields

Pilot Rain Fields, this wonderful book is about flying beyond your wildest dreams in School.

This is a great and fun book for kids of all ages.

Baby Musicain

Rain has been playing musical Instruments every since he could crawl.

Baby Race Car driver

Rain Fields racing his way to first grade.

All kids love racecars and driving.

This is just a fun and fast book for kids of all ages.

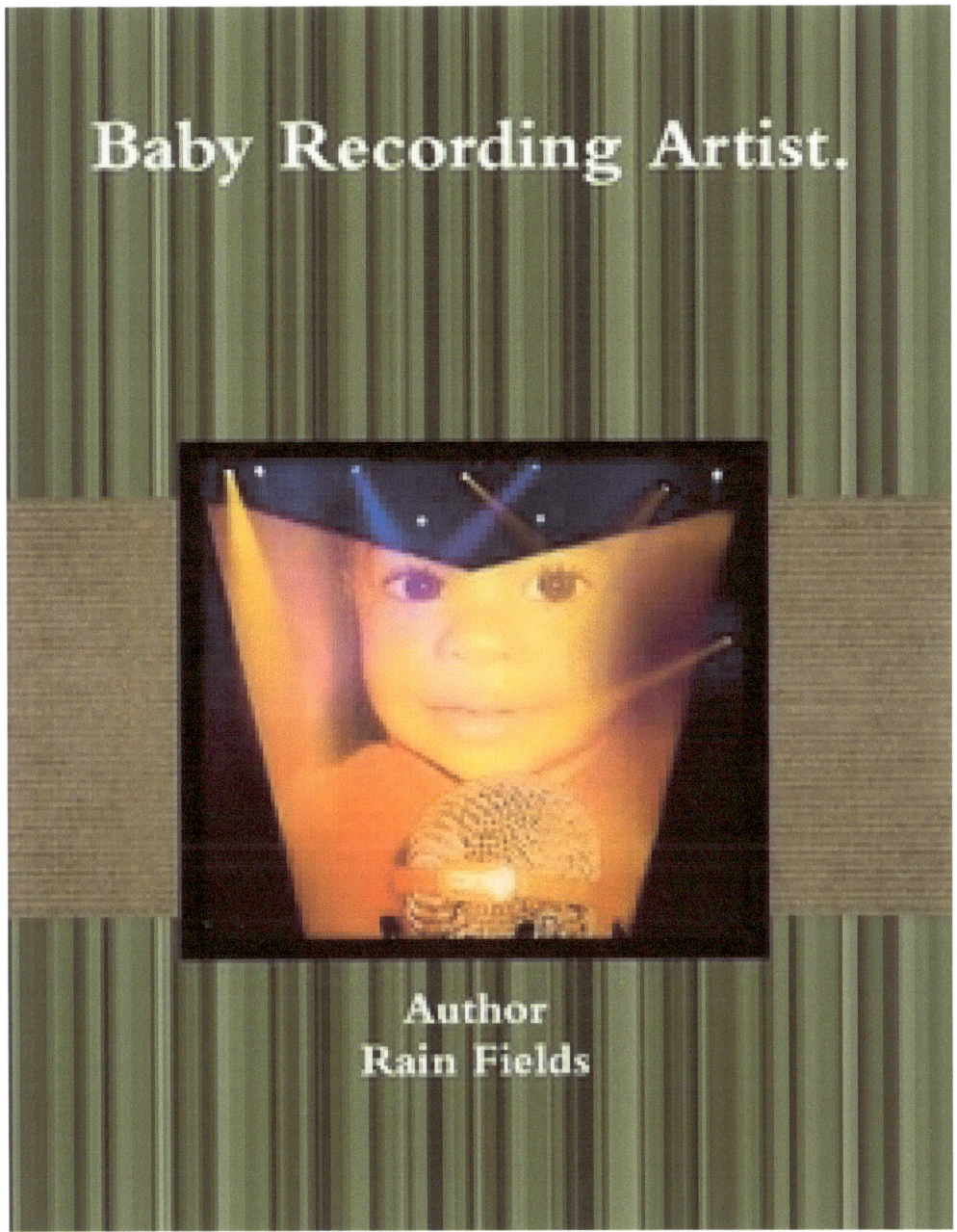

Baby Recording Artist.

Rain has big dreams of being a recording artist one day.

He has been singing and playing music every since he was just seven weeks old.

Rain loves listening to gospel and R & B music all day.

Rainbow

Rain Fields, Story book of Poems.

Rain has enjoyed, creating wonderful fun poems.

The storybook of poems, are a lot of fun.

These are great poems, for kids of all ages.

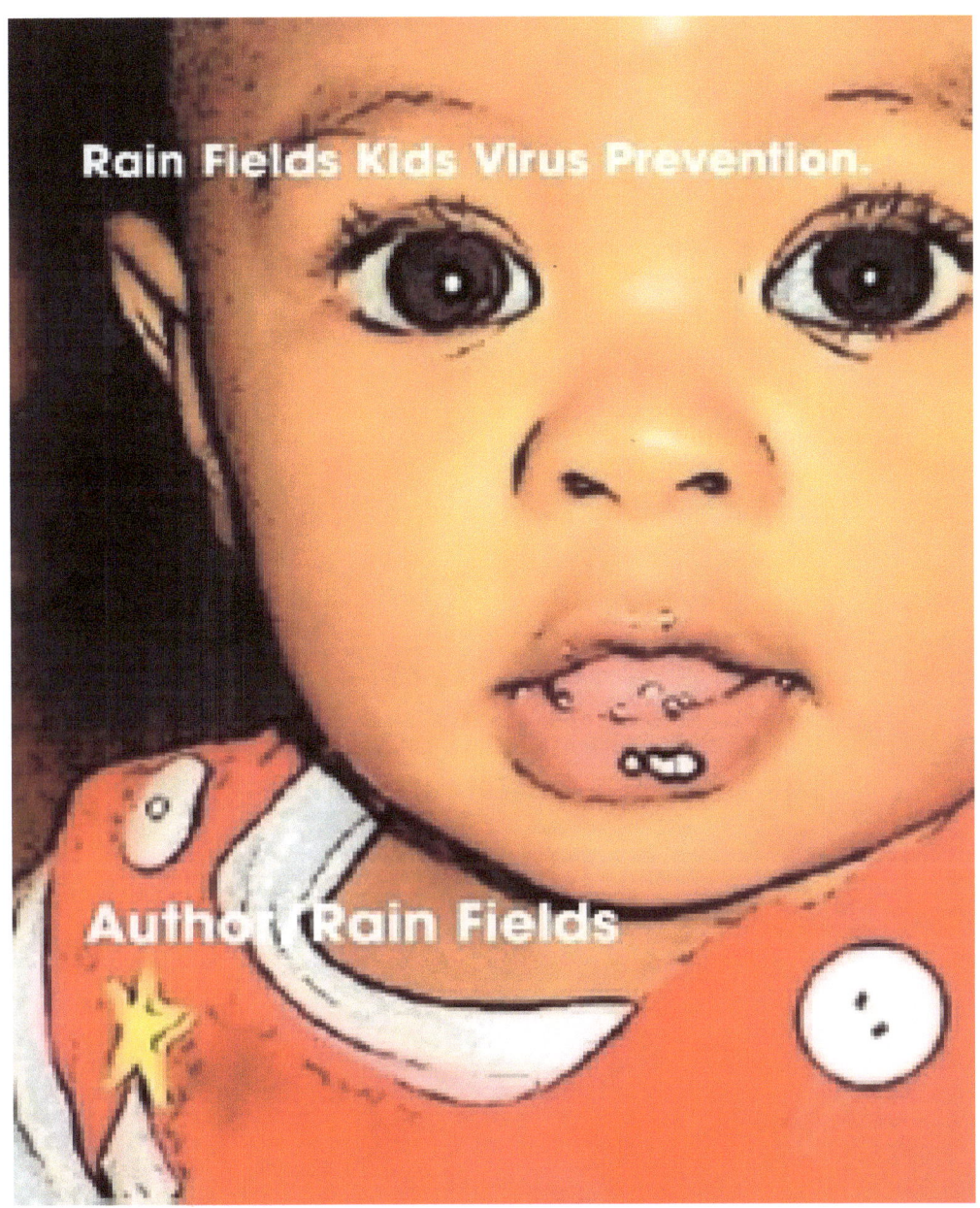

Rain Fields Kids Virus Prevention, is a new product that has been invented by me to protect against germs and bacteria.

Kids Virus Prevention is a raincoat from germs and flu like virus from kids at pre-school, day care, public restrooms.

Super Rain

America's favorite storybook character "Rain Fields".
Super Rain is a 4-year-old sup er hero; this character is base on a real life boy name Rain Fields.

Every time it Rain's, Rain Fields gets his super powers to help other kids in need in his community.
This character is to remind, every child in the world, that they are all super heroes at any age.
Every time you help out another child, in need you gain super strength from God himself.

Rain's Magical library.

This is a wonderful kids fairytale novel.

Its about a little boy who dreams of having his own library inside of his playroom.

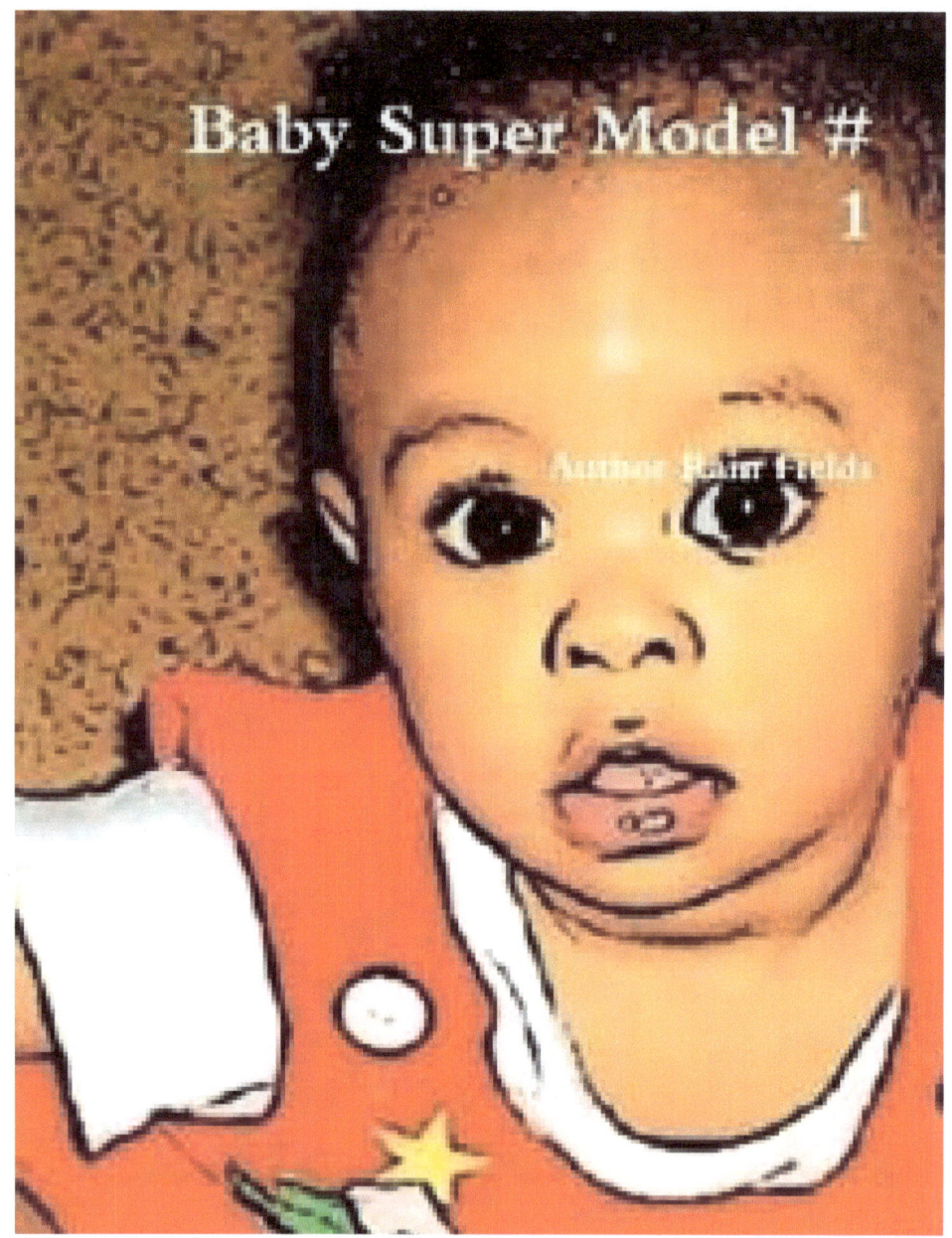

Baby Super Model

Rain loves to put on his own clothes and shoes.

Mother will make sure there is money for my education.

Rain scholarship Fund for college, money for me to attend College.

Baby Super Model #2

Rain loves to wear clothes of bright colors, with matching shoes.

He enjoys being a baby model, for kids clothes and fashion on a budget.

Rain Fields Museum

The Rain Fields Museum, will show case all of Rain books that has been published all across the globe.

It will also feature all the merchandise and products has been motivated by him.

The Rain Fields Museum for kids, will prove that hard work, does pay off.

Surfboarder Rain Fields

Rain loves to ride the waves.

He wants to go snow boarding and surfing on his next family vacation.

Sir Knight Rain Fields

There is a great chance through our family genealogy.

Rain Blood line has about 6 Knights in his family tree.

This book is to encourage children to, research their family history and tree.

Rain Fields Inc.

Rain Fields is an up and coming designer, his new line of products are for the new generation of kids.

Rain Fields has products in four categories.

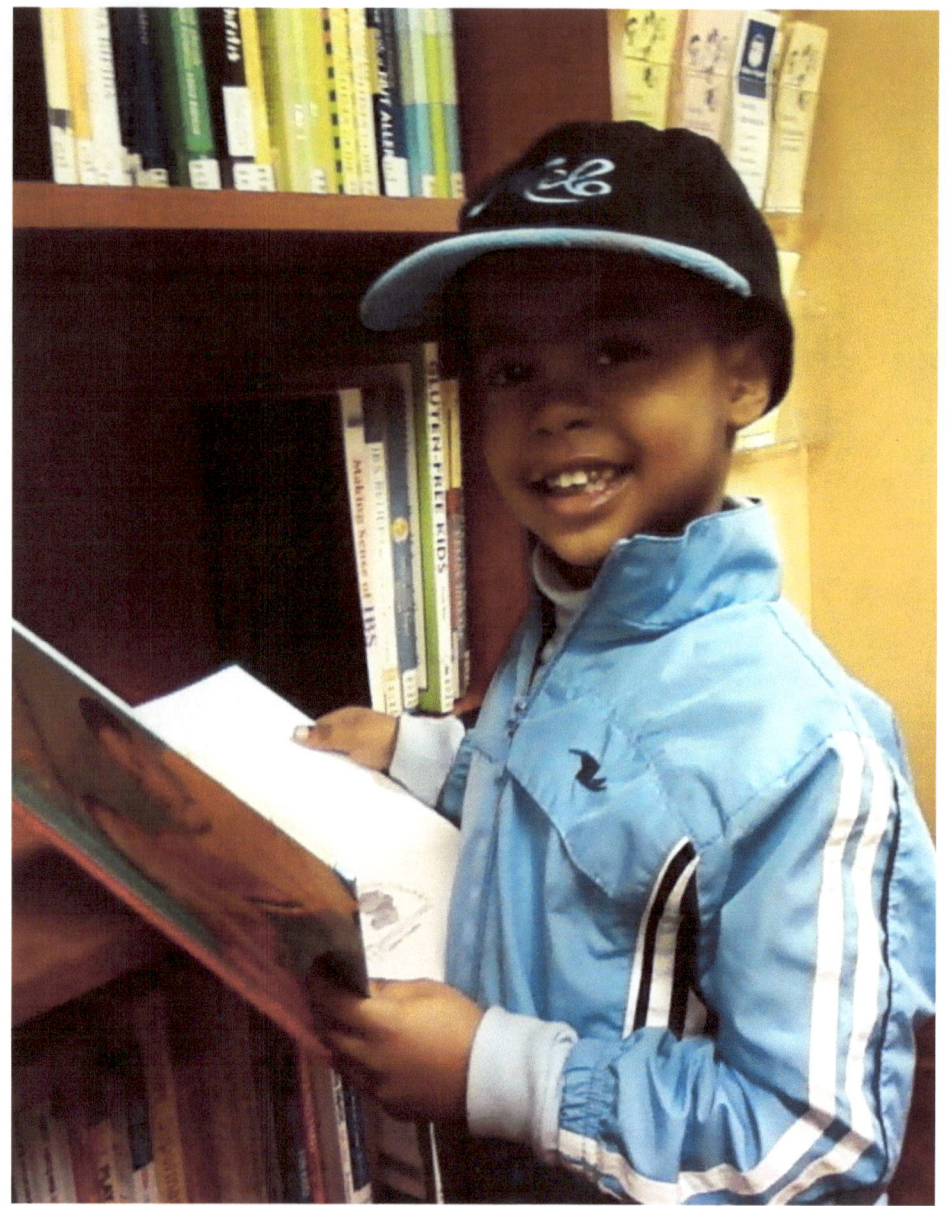

Rain Fields Children's Library.

The Rain Fields Children's Library will open soon.

This wonderful children's library, will feature all the books.

Rain has published and written with the support of his mom.

The Rain Fields Children's Library will be for kids books and Kids Magazines.

There will be story times and story readings from selected books by Rain.

This library will be for kids & children from ages 0 – 12 years old.

Rain Fields Famous friends.

Meet Rains famous friends.

Rain's Family Vacation

Rain loves traveling at fun places just for kids, with his mommy.

He wants to travel 1st class in an airplane on his next vacation.

Rain Retirement Party

Rains has been working, every since he was just six weeks old.

He plans to retire on his eighth birthday in May.

He will now focus on his school, to attend college in 10 years.

He has successfully published over 50 books,with the support and help of his mommy.

About the author

Multi-published author, Dionne Fields, who wrote the true crime series called: Pages of Me.

She is also the Director of her very own movies on paper at her Movies on Paper Studio Production.

Dionne Fields is an international children's book author with over 50 titles.

Her children's books are based on America's favorite character: Rain Fields

http://www.facebook.com/pages/Movies-On-Paper/129284647093098

I began writing novels, after publishing a dozen short stories.

Movies On Paper Studio, coming soon to a book shelf near you.

When books & movies meet at the Box Office.

I enjoy business traveling, shopping for sales and playing golf.

And family vacations with my sons

Part 1

The movies on paper children's books have a series of over 50 children's books base on the real life character Rain Fields.

The little boy who, dreamed of the impossible at any age.

This children's book collection, will inspire kids of all ages to become anything they want to be, just dream and read.

The key to life is through reading a book, and finishing school.

America's favorite storybook character "Rain Fields".

Rain Fields 1st day of work.

1. Attorney Rain – book
2. Actor Rain - book
3. Author Rain– book
4. Business Mogul Rain - book
5. Sir Knight Rain – book
6. Surfboarder Rain – book
7. President Rain -book
8. Super Model Rain - book
9. Philanthropist Rain - book
10. Fashion Model Rain- book
11. Race Car Driver Rain – book
12. Rain 1st Christmas - book

Part 2

13. CEO Rain- book
14. Rain children's book library- book
15. Rain Fields Inc- book
16. Rain Magical Library
17. Inventor Rain - book
18. Recording Artist Rain - book
19. Bully Proof Rain - book
20. Pilot Rain - book
21. Football Team Rain – book
22. Super Rain Adventure – book
23. Poet Rain– book
24. Rain Storybook of poems – book
25. Rain Retirement Party- book
26. Fireman Rain – book
27. Prince Rain – Book
28. musician Rain – book
29. Cupcakes By Rain – book
30. Astronaunt Rain – book
31. Photographer Rain- book
32. Chef Rain – book
33. Rain Family Vacation – book
34. Rain Famous Friends- book
35. God Childs – book
36. Rain Fields Twitter – https://twitter.com/d21mond
37. Rain Fields Museum – www.rainfields.blogspot.com
38. Rain Fields Company – www.rfields.webs.com
39. Rain Fields Foundation- www.rainfields.webs.com
40. Rain Fields Face book – https://www.facebook.com/pages/Rain-Fields/161250363947511

Just to name a few.

The End

www.ingramcontent.com/pod-product-compliance
Lightning Source LLC
Chambersburg PA
CBHW050722180526
45159CB00003B/1107